AI

THE BENEFITS AND RISKS OF A NEW ERA

A must-read for anyone interested in the future of technology and its impact on society.

Preface

Artificial intelligence (AI) is no longer just a buzzword; it has become an integral part of our lives. From personal assistants like Siri and Alexa to self-driving cars, AI is transforming the way we live and work. While AI has the potential to bring immense benefits to society, it also poses significant risks. As AI continues to evolve, it is crucial to explore its potential benefits and risks to make informed decisions about its development and use.

This book is a comprehensive guide to the benefits and risks of AI. In this book, we explore the ways AI is transforming various industries, including healthcare, transportation, education, and the workplace. We also discuss the potential risks associated with AI, such as job displacement, bias, and privacy concerns. Additionally, we delve into the ethical considerations of AI and the need for responsible development and use.

Through this book, we hope to provide readers with a clear understanding of the potential benefits and risks of AI and inspire them to engage in thoughtful discussions and decisions about its development and use. We hope this book serves as a valuable resource for anyone interested in the future of technology and its impact on society.

Thank you for taking the time to read this book.

TABLE OF CONTENTS

Chapter 1: Introduction
- Definition of AI
- Brief history of AI
- Importance of studying AI

Chapter 2: Benefits of AI
- AI in healthcare
- AI in transportation
- AI in education
- AI in the workplace

Chapter 3: Risks of AI
- Job displacement
- Bias and discrimination
- Autonomous weapons
- Privacy and security concerns

Chapter 4: Ethics of AI
- Ethical considerations in AI development and use
- AI and human values
- AI and social responsibility

Chapter 5: Future of AI
- Emerging trends in AI
- Opportunities and challenges
- Impact of AI on society and the economy

Chapter 6: Conclusion
- Summary of key points
- Reflection on the benefits and risks of AI
- Call to action for responsible AI development and use.

CHAPTER 1

Introduction

Artificial Intelligence (AI) is a rapidly evolving field that has been making headlines for its potential to revolutionize various aspects of our lives. From healthcare to transportation, education, and the workplace, AI has already shown tremendous promise in enhancing productivity, improving safety, and transforming our daily routines. However, as with any technology, AI also poses certain risks and ethical considerations that must be taken into account. In this chapter, we will provide an overview of AI, its definition, brief history, and the importance of studying it.

Definition of AI

AI is a branch of computer science that involves the development of intelligent machines that can perform tasks that typically require human intelligence, such as visual perception, speech recognition, decision-making, and language translation. AI can be divided into two main categories: narrow or weak AI, and general or strong AI. Narrow or weak AI is designed to perform a specific task or set of tasks, such as facial recognition or voice assistants. It is limited to the specific task it was programmed for and cannot go beyond that task. In contrast, general or strong AI aims to create machines that can perform any intellectual task that a human can. This type of AI is still in its early stages of development and is not yet a reality.

Brief History of AI

The concept of AI dates back to the 1950s when computer scientist John McCarthy coined the term "artificial intelligence." The idea of creating machines that could perform human-like tasks captured the imagination of researchers and scientists, leading to significant advancements in the field.

One of the earliest examples of AI was the game of chess. In 1997, IBM's Deep Blue defeated world chess champion Garry Kasparov in a historic match, demonstrating the potential of AI in defeating human intelligence in a game of strategy.

In recent years, advancements in machine learning and neural networks have led to significant breakthroughs in AI. For example, in 2012, a deep learning algorithm called AlexNet won the ImageNet competition, demonstrating the potential of AI in image recognition. Since then, AI has made significant strides in speech recognition, natural language processing, and robotics, among other areas.

Importance of Studying AI

Studying AI is crucial for several reasons. First, AI has the potential to transform various aspects of our lives, from healthcare to transportation, education, and the workplace. Understanding how AI works and its potential applications can help us leverage its benefits while also mitigating its risks.

Second, AI is a rapidly evolving field, and keeping up with its advancements requires ongoing learning and development.

The pace of AI development is accelerating, and it is essential to stay informed about the latest trends and emerging technologies.

Third, studying AI can help us address ethical considerations and social issues related to AI development and use. As AI becomes more integrated into our lives, it is important to ensure that it is developed and used responsibly, with due consideration for human values, privacy, and security.

Finally, studying AI can open up exciting career opportunities in a variety of fields, from software engineering and data science to robotics and healthcare.

Conclusion

In conclusion, AI is a rapidly evolving field that has the potential to transform various aspects of our lives. It is important to understand its definition, brief history, and importance to stay informed about its potential applications and mitigate its risks. As we move forward into the future, the study of AI will become increasingly important in addressing ethical considerations and addressing social issues related to AI development and use.

CHAPTER 2

Benefits of AI

Artificial Intelligence (AI) has already shown tremendous promise in enhancing productivity, improving safety, and transforming various aspects of our lives. In this chapter, we will explore some of the key benefits of AI in healthcare, transportation, education, and the workplace.

AI in Healthcare

AI has the potential to revolutionize healthcare in several ways. One of the most significant applications of AI in healthcare is the analysis of medical images, such as X-rays and MRIs. AI algorithms can analyze these images much faster and more accurately than humans, leading to earlier and more accurate diagnoses. This can be especially beneficial in rural or remote areas where access to specialized medical professionals is limited.

AI can also be used to develop personalized treatment plans. By analyzing patient data, such as medical history and genetic information, AI algorithms can identify the most effective treatments for individual patients. This can lead to better outcomes and reduced healthcare costs.

AI can also improve the efficiency of healthcare delivery by automating administrative tasks such as appointment scheduling and billing. This frees up healthcare professionals to focus on patient care and can reduce wait times and administrative errors.

AI in Transportation

AI has the potential to make transportation safer, more efficient, and more sustainable. One of the most significant applications of AI in transportation is autonomous vehicles. Self-driving cars and trucks have the potential to reduce accidents caused by human error, increase fuel efficiency, and reduce traffic congestion.

AI can also be used to optimize transportation routes and schedules.

By analyzing data such as traffic patterns and weather conditions, AI algorithms can identify the most efficient routes and schedules for public transportation, leading to reduced wait times and improved service.

AI can also enhance the safety and efficiency of air travel. For example, AI can be used to optimize flight paths, improve air traffic control, and predict and prevent equipment failures.

AI in Education

AI has the potential to transform education by providing personalized learning experiences tailored to individual students' needs. AI algorithms can analyze student data, such as learning style and progress, to develop customized lesson plans and activities. This can lead to improved learning outcomes and reduced dropout rates.

AI can also be used to automate administrative tasks such as grading and scheduling, freeing up teachers to focus on teaching and providing support to students.

AI can also enhance accessibility to education for students with disabilities.

For example, AI-powered assistive technologies such as text-to-speech and speech-to-text can enable students with hearing or vision impairments to participate fully in classroom activities.

AI in the Workplace

AI has the potential to improve productivity, reduce costs, and enhance safety in the workplace. One of the most significant applications of AI in the workplace is automation. AI-powered robots can perform repetitive or dangerous tasks, such as assembly line work or hazardous waste disposal, freeing up human workers to focus on more complex and creative tasks.

AI can also be used to optimize workflows and processes, leading to improved efficiency and reduced costs. For example, AI algorithms can analyze data to identify areas for process improvement and automate tasks such as inventory management.

AI can also enhance workplace safety. For example, AI-powered sensors can monitor workplace environments for hazards such as gas leaks or chemical spills and alert workers to potential dangers.

Conclusion

In conclusion, AI has the potential to revolutionize various aspects of our lives, from healthcare to transportation, education, and the workplace. By analyzing data and automating tasks, AI can improve efficiency, reduce costs, and enhance safety. AI also has the potential to provide personalized experiences tailored to individual needs, leading to improved outcomes and enhanced accessibility. As we move forward, it is essential to continue exploring the potential benefits of AI and leveraging them for the betterment of society.

CHAPTER 3

Risks of AI

While AI has the potential to bring significant benefits, it also presents several risks that must be addressed. In this chapter, we will explore some of the key risks of AI, including the potential for biased algorithms, job displacement, and the misuse of AI for harmful purposes.

Biased Algorithms

One of the most significant risks of AI is the potential for biased algorithms. AI algorithms are only as unbiased as the data they are trained on, and if the data is biased, the algorithm will be biased as well. This can have significant consequences, especially in applications such as hiring, lending, and criminal justice.

For example, an AI algorithm used to screen job candidates may be trained on historical data that reflects biases against certain groups. As a result, the algorithm may perpetuate those biases, leading to unfair hiring practices. Similarly, an AI algorithm used to determine whether to grant a loan may be biased against certain groups, leading to discrimination. To address this risk, it is essential to ensure that AI algorithms are trained on unbiased data and are regularly audited to detect and correct any biases that may arise.

Job Displacement

Another significant risk of AI is the potential for job displacement. As AI algorithms become more sophisticated, they are increasingly capable of performing tasks that were previously performed by humans. This could lead to significant job losses in certain industries, such as manufacturing and transportation. However, it is important to note that AI also has the potential to create new jobs and industries. For example, the development and deployment of AI algorithms require specialized skills, and the growth of the AI industry could lead to the creation of new jobs in areas such as data science and software development.

To address the risk of job displacement, it is important to invest in retraining and reskilling programs to prepare workers for the jobs of the future.

Misuse of AI

Another significant risk of AI is the potential for misuse, particularly for harmful purposes. For example, AI could be used to develop autonomous weapons or to automate surveillance systems, leading to significant violations of human rights.

AI could also be used to perpetuate disinformation campaigns, leading to the spread of false information and the manipulation of public opinion.

To address the risk of AI misuse, it is important to establish ethical guidelines for the development and deployment of AI systems. This includes ensuring that AI is used in ways that align with human values and promoting transparency and accountability in AI decision-making processes.

Conclusion

In conclusion, while AI presents significant benefits, it also presents significant risks that must be addressed. Biased algorithms, job displacement, and the misuse of AI for harmful purposes are all significant concerns that require careful consideration. To ensure that AI is used for the betterment of society, it is essential to develop and implement ethical guidelines for the development and deployment of AI systems. By doing so, we can ensure that AI continues to bring benefits while minimizing the risks.

CHAPTER 4

Ethics of AI

As AI continues to advance and become increasingly integrated into our lives, it is essential to consider the ethical implications of this technology. In this chapter, we will explore some of the key ethical considerations surrounding AI, including transparency, accountability, and privacy.

Transparency

Transparency is a critical ethical consideration when it comes to AI. AI algorithms often operate as black boxes, making it difficult for users to understand how decisions are being made. This lack of transparency can make it challenging to detect and correct biases or errors in AI systems. To address this, it is essential to promote transparency in AI decision-making processes.

This can include providing explanations for AI decisions, making AI algorithms open source, and establishing auditing processes to ensure that AI systems are operating in an ethical and transparent manner.

Accountability

Another critical ethical consideration is accountability. As AI becomes more integrated into our lives, it is essential to ensure that those responsible for developing and deploying AI systems are held accountable for their actions. This includes ensuring that AI systems are designed to minimize harm, that data is collected and used ethically, and that users are protected from discrimination and other forms of harm. To promote accountability, it is essential to establish clear regulations and guidelines for the development and deployment of AI systems. This can include establishing ethical standards for AI developers and requiring transparency in AI decision-making processes.

Privacy

Privacy is another critical ethical consideration when it comes to AI. AI systems often collect large amounts of data on individuals, which can be used to make decisions about them.

This can lead to significant privacy concerns, particularly if the data is used in ways that individuals do not expect or that violate their rights.

To address these concerns, it is essential to establish clear guidelines for the collection and use of data in AI systems. This can include requiring informed consent from individuals, limiting the collection and use of personal data, and ensuring that data is used in ways that are consistent with ethical principles.

Conclusion

In conclusion, the ethical considerations surrounding AI are complex and multifaceted. Transparency, accountability, and privacy are just a few of the critical ethical considerations that must be addressed as AI continues to advance and become increasingly integrated into our lives. To ensure that AI is used in ways that align with human values and promote the betterment of society, it is essential to establish clear regulations and guidelines for the development and deployment of AI systems. By doing so, we can promote ethical AI development and ensure that this technology is used to benefit humanity.

CHAPTER 5

Future of AI

As AI continues to advance at a rapid pace, it is essential to consider the future of this technology and the potential impact it will have on society. In this chapter, we will explore some of the potential developments and applications of AI in the future.

Advanced Automation

One of the most significant impacts of AI in the future will be the continued automation of tasks that were previously done by humans. As AI systems become increasingly advanced and capable, they will be able to take on more complex tasks, from driving cars to diagnosing medical conditions.

This automation has the potential to revolutionize industries and create significant efficiencies. However, it also raises concerns about the potential loss of jobs and the impact on the workforce. To address these concerns, it will be essential to focus on retraining and reskilling workers, as well as developing policies and regulations to ensure that the benefits of automation are shared fairly across society.

AI-Powered Healthcare

Another area where AI is likely to have a significant impact is healthcare. AI systems have already shown great promise in areas such as medical imaging and drug discovery. In the future, these systems could be used to develop more personalized treatments and to identify potential health risks before they become serious problems.

However, the use of AI in healthcare also raises concerns about privacy and the potential for discrimination. To address these concerns, it will be essential to establish clear regulations and guidelines for the collection and use of healthcare data, as well as ensuring that AI systems are developed and used in ways that prioritize patient privacy and well-being.

Intelligent Virtual Assistants

One of the most exciting potential developments in AI is the use of intelligent virtual assistants. These systems, such as Amazon's Alexa and Apple's Siri, are already in widespread use and are becoming increasingly advanced.

In the future, these systems could be used for a wide range of tasks, from booking appointments to providing personalized recommendations. However, they also raise concerns about privacy and the potential for these systems to become too intrusive in our daily lives. To address these concerns, it will be essential to establish clear guidelines for the collection and use of personal data and to ensure that users have control over how their data is used.

Conclusion

In conclusion, the future of AI is both exciting and uncertain. The continued automation of tasks, the use of AI in healthcare, and the development of intelligent virtual assistants are just a few of the potential developments that we can expect to see in the coming years.

However, as with any technology, AI also raises concerns about the potential impact on society, from job loss to privacy concerns. To ensure that AI is used in ways that promote the betterment of humanity, it will be essential to establish clear regulations and guidelines for the development and deployment of AI systems. By doing so, we can ensure that AI is used in ways that align with human values and promote the betterment of society.

CHAPTER 6

Conclusion

In this book, we have explored the benefits and risks of artificial intelligence and the importance of ethical considerations in the development and deployment of AI systems. We have seen how AI is already being used to revolutionize industries such as healthcare, transportation, and education, and how it has the potential to create significant efficiencies and improve the quality of life for people around the world.

However, we have also seen that AI is not without risks. The continued automation of tasks raises concerns about the impact on the workforce and the potential for job loss. The use of AI in healthcare raises concerns about privacy and the potential for discrimination.

And the development of intelligent virtual assistants raises concerns about the potential for these systems to become too intrusive in our daily lives.

To address these risks, it is essential to prioritize the ethical development and deployment of AI systems. This includes establishing clear regulations and guidelines for the collection and use of personal data, as well as ensuring that AI systems are developed and used in ways that prioritize human values and promote the betterment of society.

As AI continues to advance, it will be essential to remain vigilant and proactive in addressing the potential risks and ensuring that the benefits of this technology are shared fairly across society. By doing so, we can help to ensure that AI is used in ways that promote the betterment of humanity and create a brighter future for all.

www.ingramcontent.com/pod-product-compliance
Lightning Source LLC
Chambersburg PA
CBHW040350220526
45473CB00009B/2836